Stepping into Standards Theme Series

Weather Wonders

Written by

Kimberly Jordano and Tebra Corcoran

Editors: Carla Hamaguchi and Teri L. Fisch
Illustrator: Jenny Campbell
Cover Illustrator: Kimberly Schamber
Designer: Moonhee Pak
Cover Designer: Moonhee Pak
Art Director: Tom Cochrane
Project Director: Carolea Williams

Table of Contents

Introduction

Due to the often-changing national, state, and district standards, it is often difficult to "squeeze in" fascinating topics for student enrichment on top of meeting required standards and including a balanced program in your classroom curriculum. The *Stepping into Standards Theme Series* will help you incorporate required subjects and skills for your kindergarten and first-grade children while engaging them in a fun theme. Children will participate in a variety of language arts experiences to help them with **phonemic awareness** and **reading** and **writing** skills. They will also have fun with **math activities, hands-on science activities,** and **social studies class projects.**

The creative lessons in *Weather Wonders* provide imaginative, innovative ideas to help you motivate children as you create the four seasons in your classroom. The activities will inspire children to explore weather and seasons as well as provide them with opportunities to enhance their knowledge and meet standards.

Invite children to explore weather as they

- participate in phonemic awareness activities that feature theme-related poems and songs
- create mini-books that reinforce guided reading and sight word practice
- contribute to shared and independent reading and writing experiences about different types of weather
- practice sorting, counting, and adding with rainbow treats
- learn about the different types of clouds
- discuss how the different weather types make them feel
- complete several fun art projects as they create the four seasons in their classroom
- participate in an end-of-the-unit Weather Wonderland to showcase their work

Each resource book in the *Stepping into Standards Theme Series* includes standards information, easy-to-use reproducibles, and a full-color overhead transparency to help you integrate a fun theme into your required curriculum. You will see how easy it can be to incorporate creative activities with academic requirements while children enjoy their exploration of weather!

Meeting Standards

Language Arts	It's a Sunny Day, page 7	The Seasons Song, page 7	Weather Song, page 8	Yummy Hot Chocolate, page 8	Reading Aloud, page 15	Morning Message, page 16	Pocket Chart Stories, page 17	Sentence Puzzle, page 17	Assembling the Reading Sticks, page 19	What's the Weather Like Today? mini-book, page 20	Look! A Rainbow! mini-book, page 24	Windy Windsock Bag, page 26	What's the Weather?, page 26	Umbrella Word Wall, page 30	Weather Wizard of the Day, page 31	Weather Facts, page 34	Where Is Little Mouse?, page 34	Weather Warm-Up, page 35	The Snowy Day, page 37	Clouds, page 38	Have You Ever Built a Snowman?, page 39
Phonemic Awareness																					
Blend vowel-consonant sounds to make words	•																				•
Identify rhyming words		•		•									•								
Distinguish letters from words			•			•		•	•	•				•		•	•	•			
Distinguish beginning consonant sounds				•										•		•	•	•			
Improve oral language	•	•	•	•								•	•		•				•	•	
Reading																					
Concepts about print					•	•	•	•	•	•	•	•				•	•	•	•	•	•
Improve reading comprehension					•							•	•				•		•	•	
Track text			•			•	•	•	•	•	•					•	•	•		•	•
Improve reading fluency			•			•	•	•		•	•	•	•			•					
Identify structural features of informational text					•							•									
Make predictions about story content					•							•									
Identify story elements					•														•	•	
Recognize high-frequency and sight words						•	•	•	•	•	•			•			•				•
Writing																					
Brainstorm and organize ideas														•		•			•	•	
Understand simple sentence structure								•		•	•				•	•	•	•	•	•	•
Develop focused, detailed writing																•	•		•	•	
Use inventive spelling																•	•		•	•	•
Spell words independently																	•	•	•	•	
Write complete sentences										•	•				•		•		•		•
Choose correct punctuation																•	•		•	•	•
Incorporate letter formation and word spacing										•				•		•	•				•

Meeting Standards

Math Science Social Studies	Edible Rainbows, page 40	Rainbow Colors, page 40	Seasons Glyph Quilt, page 41	Winter Ice Cube Counting, page 42	Sunshine Solar Sandwiches, page 48	Rainy Day in a Bag, page 48	The Wind Is Blowing, page 49	My Big Yellow Boots, page 52	Feelings for Weather, page 53	Weather Wonderland Invitation, page 57	Snow Globes, page 57	Create a Weather Wonderland, page 58	Cloud Viewers, page 58	Windy Day, page 59	Springtime Rainbow Bubbles, page 60	Bubble Paint, page 60	Cooking, page 60
Math																	
Sort	●																
Count	●	●	●														●
Add	●																
Subtract	●																
Graph		●															
Analyze data		●	●														
Recognize numbers				●													
Write numbers	●			●													
Compare (greater, less)		●															
Science																	
Experiment and observe					●	●	●						●				
Concept of melting					●												
Prediction and hypothesis						●											
Condensation						●											
Wind and weather					●		●				●	●		●	●		
Types of clouds													●				
Chemistry—make bubble solution															●	●	
Social Studies																	
Manners								●		●							
Respect								●									
Feelings									●								
Additional Language Arts																	
Improve reading fluency								●									●
Reading informational text										●							●
Use inventive spelling								●		●				●			
Understand simple sentence structure								●	●					●			

Instant Learning Environment

This resource includes a full-color overhead transparency of weather in the four seasons that can be used in a variety of ways to enhance the overall theme of the unit and make learning more interactive. Simply place the transparency on an overhead projector, and shine it against a blank wall, white butcher paper, or a white sheet. Then, choose an idea from the list below, or create your own ideas for using this colorful backdrop.

Unit Introduction
Give children clues about the weather unit. For example, say *We are going to study about temperature, rain, and seasons*. Invite children to use the clues to predict and discuss what the unit might be about. Then, display the transparency to give children a quick overview of the environment and an introduction to the unit.

Or, cut out puzzle pieces from an 8½" x 11" (21.5 cm x 28 cm) sheet of paper. Place the puzzle pieces on top of the transparency on the overhead projector so they cover it entirely. Turn on the projector. None of the transparency will show. Remove one puzzle piece at a time, and describe the uncovered section. Invite children to identify the scene. Then, continue to remove pieces, asking children to predict what they might see next until you have revealed the entire transparency.

Dramatic Play
Use the transparency as a backdrop for children to perform the dramatic play described on page 26. Have children hold a puppet (glue cutouts from the Weather Items reproducible on page 29 to craft sticks) or hold up weather props (e.g., umbrella, sunglasses) as they perform.

Here are some other uses for the transparency:

- Use the transparency as a backdrop for children to do an oral presentation (e.g., My favorite weather is _____).

- Use the transparency as part of an interactive activity. Have children use a pointer to point to and identify the various types of weather and seasons. Invite children to point to the tree that matches the clothes they are wearing. For example, a child who is wearing shorts and a T-shirt would point to the summer tree.

Children listen to and identify the sounds that make up words and manipulate sounds to create words.

Phonemic Awareness

It's a Sunny Day

Copy, color, and cut apart a set of the Sunny Day Picture Cards. Display the picture cards on the board or in a pocket chart. Sing "It's a Sunny Day!" As children correctly blend the words (e.g., /b/ /ike/), show the matching picture. Or, have a child find the correct picture and show it to the class. To extend the activity, use the two extra picture cards to add verses to the song using the onsets and rhymes for /j/ /ump/ and /sk/ /ate/.

MATERIALS

- ✓ Sunny Day Picture Cards (page 9)
- ✓ "It's a Sunny Day!" song (page 11)
- ✓ pocket chart (optional)

The Seasons Song

Make several copies of the Season Characters reproducible. Color and cut apart each item. Glue the items to separate tongue depressors to create puppets. Give each child a puppet, and sing the song. Invite children to march around the classroom and hold up their puppet at the appropriate part of the song (i.e., snowman for winter, flower face for spring, sunshine face for summer, squirrel for fall).

MATERIALS

- ✓ Season Characters reproducible (page 10)
- ✓ "The Seasons" song (page 11)
- ✓ tongue depressors

Weather Song

MATERIALS

- ✓"Weather" song (page 12)
- ✓chart paper
- ✓tagboard
- ✓Velcro®

Write each verse of the song "Weather" on a separate piece of chart paper. Write one of the spelled-out weather words (e.g., windy) on tagboard. Laminate the word card, and cut apart the word. Use Velcro to attach each letter card to the chart. Use markers and construction paper to decorate the chart. Use a reading stick (see page 19) to track the words as they are sung. Place each letter card on the chart as that part of the song is sung. For example, when the class sings, "And windy was the weather! W-I," you place the W and I cards on the chart and children clap their hands and omit saying the letters N, D, and Y.

Yummy Hot Chocolate

MATERIALS

- ✓Hot Chocolate reproducible (page 13)
- ✓Marshmallow Cards (page 14)
- ✓small paper bags

Give each child a Hot Chocolate reproducible, a set of Marshmallow Cards, and a paper bag. Ask children to cut out the hot chocolate mug and glue it on the front of their paper bag. Have children cut apart their marshmallow cards. Then, invite them to sort their cards into two piles: words that begin with /ch/ and words that begin with /k/. Invite the class to read the poem together. Have children put the picture cards (marshmallows) that begin with /ch/ in their bag (cup of hot chocolate). Encourage children to say each /ch/ word (e.g., The **chick** goes in the chocolate) as they place the cards in their bag to further reinforce the /ch/ sound.

Sunny Day Picture Cards

Season Characters

It's a Sunny Day!

(sing to the tune of "Yankee Doodle")

I'm going to go outside today
Because there's lots of sun.
See if you can figure out what I will do for fun.

I'm going to go outside today.
Can you guess what I'll ride?
It starts with /**b**/ and ends with /**ike**/. Yes! I will **bike!**

I'm going to go outside today.
Can you guess what I'll do?
It starts with /**sw**/ and ends with /**ing**/. Yes! I will **swing!**

I'm going to go outside today.
Can you guess what I'll do?
It starts with /**cl**/ and ends with /**imb**/. Yes! I will **climb!**

I'm going to go outside today.
Can you guess what I'll do?
It starts with /**d**/ and ends with /**ig**/. Yes! I will **dig!**

The Seasons

(sing to the tune of "The Ants Go Marching")

The seasons they go round and round, hurrah, hurrah,
The seasons they go round and round, hurrah, hurrah.
Winter, spring, summer, and fall.
The weather changes with them all.
And the seasons they go round, and round, and round
and round and round!

Weather

(sing to the tune of "Bingo")

The Wind

There was a day that was breezy,
And windy was the weather!
W-i-n-d-y, w-i-n-d-y, w-i-n-d-y,
And windy was the weather!

The Sun

There was a day that was hot,
And sunny was the weather!
S-u-n-n-y, s-u-n-n-y, s-u-n-n-y,
And sunny was the weather!

The Fog

There was a day that was misty,
And foggy was the weather!
F-o-g-g-y, f-o-g-g-y, f-o-g-g-y,
And foggy was the weather!

The Rain

There was a day that was wet,
And rainy was the weather!
R-a-i-n-y, r-a-i-n-y, r-a-i-n-y,
And rainy was the weather!

The Snow

There was a day that was cold,
And snowy was the weather!
S-n-o-w-y, s-n-o-w-y, s-n-o-w-y,
And snowy was the weather!

Hot Chocolate

Marshmallow Cards

The teacher reads aloud a book to children while modeling fluency, expression, and enjoyment of text.

Modeled Reading

Introduce weather to your class by reading aloud books from the following literature list or others with similar content. Invite children to look at the book cover and pictures and discuss what they see. Ask them to predict what the book will be about and to point out details that relate to weather. Ask them to make connections to their own lives by sharing weather experiences.

Literature List

Animals in Winter by Henrietta Bancroft and Richard Van Gelder (Econo-Clad Books)
The Biggest, Best Snowman by Margery Cuyler (Scholastic)
The Cloud Book by Tomie dePaola (Holiday House)
Clouds by Roy Wandelmaier (Troll)
Cloudy with a Chance of Meatballs by Judi Barrett (Scott Foresman)
The Happy Day by Ruth Krauss (HarperCollins)
The Hat by Jan Brett (Penguin Putnam)
In the Snow: Who's Been There? by Lindsay Barrett George (Greenwillow Books)
It Looked Like Spilt Milk by Charles G. Shaw (HarperCollins)
It's Raining, It's Pouring by Kin Eagle (Charlesbridge Publishing)
The Jacket I Wear in the Snow by Shirley Neitzel (Mulberry Books)
A Little Bit of Winter by Paul Stewart (HarperCollins)
Little Cloud by Eric Carle (Philomel)
The Magic School Bus Kicks Up a Storm by Nancy White (Scholastic)
The Mitten by Jan Brett (Penguin Putnam)
Mushroom In the Rain by Mirra Ginsburg (Econo-Clad Books)
The Napping House by Audrey Wood (Harcourt)
Rain by Robert Kalan (Econo-Clad Books)
Rain by Rozanne Lanczak Williams (Creative Teaching Press)
The Rainbabies by Laura Krauss Melmed (Lothrop, Lee & Shepard)
Red Leaf, Yellow Leaf by Lois Ehlert (Harcourt)
Snow by Roy McKee and P. D. Eastman (Random House)
Snowballs by Lois Ehlert (Harcourt)
The Snowy Day by Ezra Jack Keats (Viking)
Splish, Splash, Spring by Jan Carr (Holiday House)
Summer by Alice Low (Random House)
Thunder Cake by Patricia Polacco (Scholastic)
Time to Sleep by Denise Fleming (Henry Holt and Company)
The Tiny Seed by Eric Carle (Simon & Schuster)
Weather Words and What They Mean by Gail Gibbons (Holiday House)
When Winter Comes by Nancy Van Laan (Atheneum)
The Wind Blew by Pat Hutchins (Econo-Clad Books)

The teacher and children read together from an enlarged text.

Shared Reading

Morning Message

MATERIALS

- ✓ chart paper or dry erase board
- ✓ markers or dry erase markers
- ✓ Wikki Stix® (optional)
- ✓ reading stick

Turn your morning message into a weather experience! This activity is a great way to introduce your new theme. Write a message (see sample below) on chart paper or a dry erase board each morning. As you write, invite children to help you sound out words, spell words, and decide what to write. Create a "secret code," and write a "secret message" for children to decode each day. Write the alphabet, and write a number above each letter (e.g., A–1, Z–26), or put numbers below your posted alphabet. Beneath the morning message, draw a blank and a number for each letter of the secret message (umbrella).

Dear Weather Wizards in Room ___.
Today is Monday, March ___, 200__.
Today we will learn about the rain!
Can you read my code?

__ __ __ __ __ __ __ __
21 13 2 18 5 12 12 1

1	2	3	4	5	6	7
A	B	C	D	E	F	G
8	9	10	11	12	13	14
H	I	J	K	L	M	N
15	16	17	18	19	20	21
O	P	Q	R	S	T	U
22	23	24	25	26		
V	W	X	Y	Z		

Invite children to write in the room number and date with a marker or dry erase marker. Have them circle letters, words, or punctuation with a marker, a dry erase marker, or Wikki Stix. Depending on the level of the children, leave complete words or word chunks deleted for them to fill in. Have volunteers write the missing letters in the coded message. Have children read aloud the completed message. Choose a child to be the "weatherperson of the day," and invite him or her to use a reading stick (see page 19) to track and reread the morning message.

Pocket Chart Stories

MATERIALS

- ✓ mini-book reproducibles (pages 20–23 and 24)
- ✓ sentence strips
- ✓ colored markers
- ✓ pocket chart
- ✓ sticky notes

Choose a mini-book, and write each sentence on a separate sentence strip. Highlight key words by writing them in a different color to help children easily recognize them. Place the sentence strips in a pocket chart. Make copies of the mini-book pictures, color them, and place each picture next to the matching sentence. Have the class read aloud the story while you track and stress high-frequency words. Invite the class to revisit the story. Select a word, letter, or part of a word, and cover it with a sticky note. Invite children to use reading strategies to identify the selected word. Remind them to look at the beginning or end sound and to decide if their answer makes sense in the sentence.

Sentence Puzzle

MATERIALS

- ✓ sentence strips
- ✓ pocket chart

Choose a sentence strip from the pocket chart story (see above), and cut it apart to create word cards. Pass out the cards, and have children read aloud their word. Invite children with word cards to stand up and arrange their words so they form a sentence. Have them put the cards back in the pocket chart in the correct order.

The teacher guides a group of children with similar reading abilities and similar skill needs through the same mini-book.

Guided Reading

Assembling the Mini-Books

MATERIALS

- ✓ mini-book reproducibles (pages 20–23 and 24)
- ✓ Rainbow Book Pattern (page 25)
- ✓ tagboard
- ✓ construction paper
- ✓ close-up photo of each child's face

Make single-sided copies of the reproducibles for the mini-book *What's the Weather Like Today?* Fold each page in half so the blank side of the paper does not show, and staple the pages inside a construction paper cover so that the creased sides face out. Make six copies of each child's photo. Have children cut out their photos and glue one on each page of their book. Or, have children draw their face on each page. Have children draw clothing on the child on the last page of their book.

Copy the Rainbow Book Pattern on tagboard. Trace the largest arc on red construction paper, and cut it out to make a pattern. Then, cut off the largest arc on the tagboard, trace the next arc on orange construction paper, and cut it out to make a pattern. Cut off that arc, and continue in the same manner for the remaining arcs to make a pattern for each one. Trace and cut out a class set of arcs on the appropriate colors of construction paper. Staple together a set of arcs for each child inside a white construction paper cover (the same size as the red arc) to make individual books. Give each child an assembled book and a "Look! A Rainbow!" sentence strips reproducible. Ask children to cut apart the strips and glue each one on the appropriate page of their book. Have them glue the title strip on the front cover and the strip that says ______! *I see a rainbow!* on the inside of their back cover.

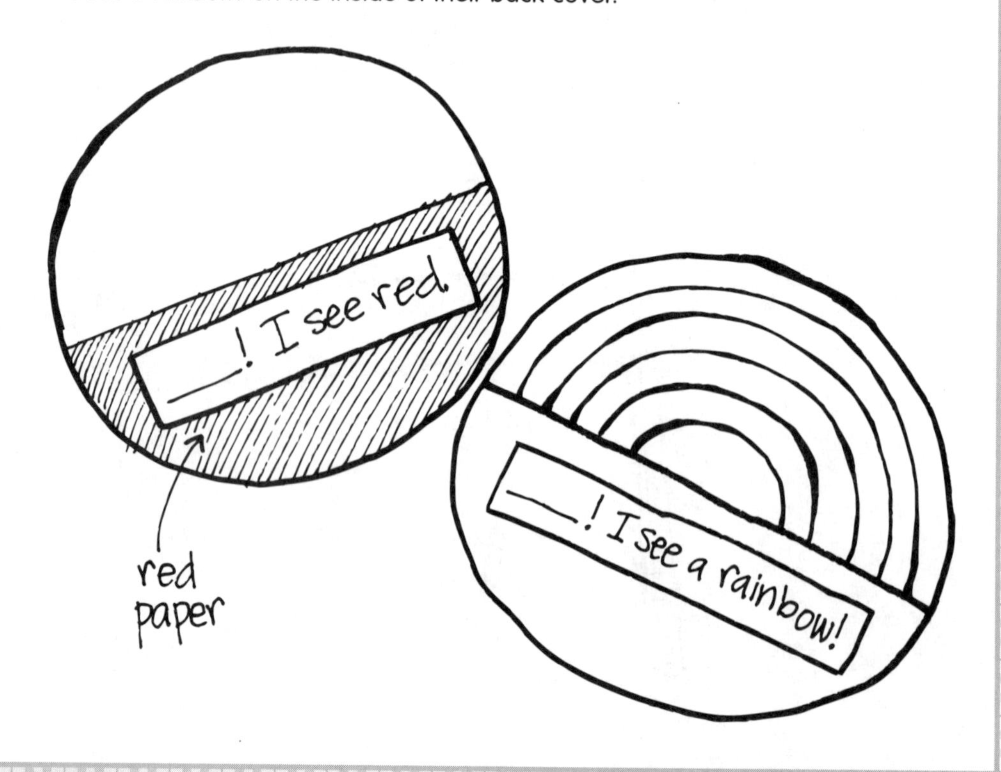

Assembling the Reading Sticks

MATERIALS

- ✓craft sticks
- ✓stickers or small objects
- ✓envelopes

Reading sticks help children with one-to-one correspondence and left-to-right directionality and are fun to use. To make a reading stick, glue to the end of a craft stick a sticker or small object that relates to the theme of the mini-book. For example, use a sun, raindrop, cocktail umbrella, or cotton ball (cloud or snowball) for *What's the Weather Like Today?* and a rainbow for *Look! A Rainbow!* Seal envelopes, and cut them in half. Glue each envelope to the front inside cover of a mini-book to make a "pocket." Place a reading stick in the pocket.

Sight Word Practice

MATERIALS

- ✓assembled mini-books (see page 18)
- ✓assembled reading sticks (see above)
- ✓art supplies

After children review the mini-book text in a shared reading lesson (see page 17), have them write the missing sight words in the blanks to complete their mini-book. In *What's the Weather Like Today?*, children will write the sight word *what* in the first blank. Ask children to look at the picture on each page for clues of what the weather is like and then write the corresponding word in the second blank (i.e., sunny, cloudy, rainy, windy, snowy). This book can be sung to the tune of "London Bridge." In *Look! A Rainbow!*, the sight word is *look*. Invite children to decorate their book cover and color the illustrations in *What's the Weather Like Today?* Have children use reading sticks to help them track words as they read the stories in guided reading groups.

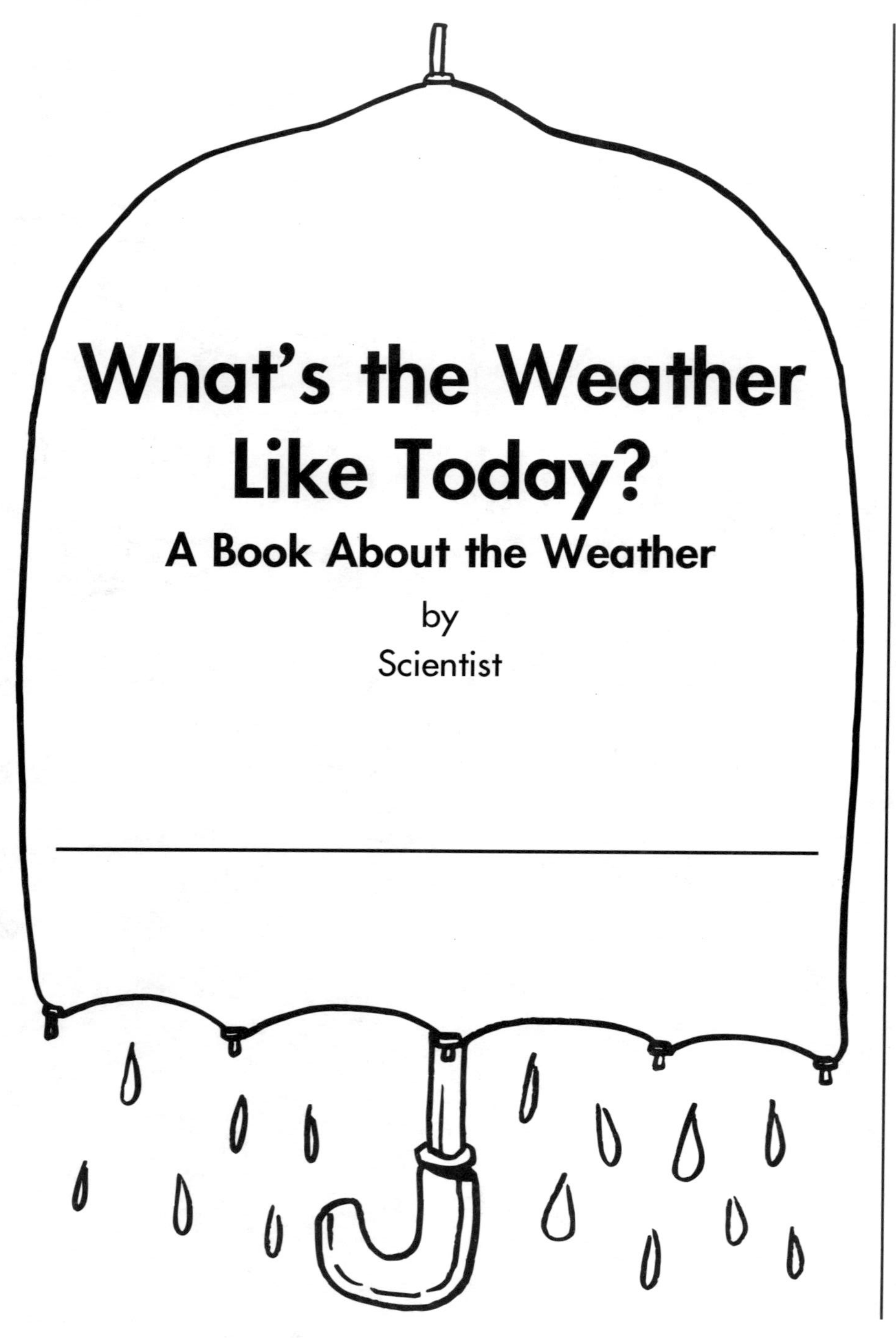

What's the Weather Like Today?

A Book About the Weather

by

Scientist

Dedicated to

____

_______ is the weather like today?

Today is ___________.

3

_______ is the weather like today?

Today is ___________.

4

_______ is the weather like today?

Today is ___________.

5

_______ is the weather like today?

Today is ___________.

6

_______ is the weather

like today?

Today is ___________.

Sun, rain, or snow!

It's time to play.

So let's go!

The End

Look! A Rainbow!

by ______________________________

____________! I see red.

____________! I see orange.

____________! I see yellow.

____________! I see green.

____________! I see blue.

____________! I see purple.

____________! I see a rainbow!

Rainbow Book Pattern

red

orange

yellow

green

blue

purple

Independent Reading

Windy Windsock Bag

MATERIALS

- ✓ books about the wind
- ✓ Windy Windsock Family Letter (page 27)
- ✓ backpack or bag
- ✓ yarn
- ✓ colored paper bags
- ✓ paper streamers
- ✓ hole punch
- ✓ resealable plastic bag with art supplies (e.g., glitter, sequins)

Copy and laminate the Windy Windsock Family Letter, and tie it to a backpack or bag. Place yarn, colored paper bags, paper streamers, a hole punch, a bag with art supplies, and books about the wind in the backpack. Tell children they will each have a turn to take home the Windy Windsock Bag and will read books about the wind and make their own windsock with their family's help. Send the backpack home with a different child each night. On the following day, have the child share his or her windsock with the class.

What's the Weather?

MATERIALS

- ✓ What's the Weather? reproducible (page 28)
- ✓ Weather Items reproducible (page 29)
- ✓ weather color transparency
- ✓ weather items (e.g., bathing suit, sweatshirt, mittens, umbrella) (optional)
- ✓ craft sticks
- ✓ overhead projector

Invite your class to perform a dramatic play. Read aloud the play on the What's the Weather? reproducible to introduce children to their lines. Assign every child a part. Give each child a cutout from the Weather Items reproducible to color. Invite children to glue their cutout to a craft stick to make a puppet. Or, give children real weather items to hold up or wear as they perform. Project the transparency onto a blank wall as the backdrop for the dramatic play. Have children practice their lines until they know them and are ready to perform in front of an audience. (This dramatic play can be read to the tune of "Zip-A-Dee-Doo-Dah.")

Windy Windsock Family Letter

Dear Family,

We have been very busy in our classroom learning about different kinds of weather! Tonight your little "meteorologist" is bringing home the Windy Windsock Bag. Have your child read the enclosed books about the wind to you or with you. Please help your child create a windsock. Begin by choosing one colored bag. Cut off the bottom of the bag and attach a yarn handle (use the hole punch). Have your child use the enclosed art supplies to decorate the windsock to make it special. Add streamers to the bottom of it. If you would like to decorate the windsock with any other fun materials, we would appreciate it!

Please return all of the items in the bag for another child to use tomorrow. Don't forget to have your child bring his or her windsock to school tomorrow to share with the class.

Thank you for your help!

Sincerely,

What's the Weather?

All: What's the weather like today? Let's ask the children and see what they say!

Bathing suits: *(hold up bathing suits)* We'll wear our bathing suits out today! It looks like it's going to be a sunny day!

All: What's the weather like today? Let's ask the children and see what they say!

Sweatshirts: *(hold up sweatshirts)* We'll wear our sweatshirts out today! It looks like it's going to be a windy day!

All: What's the weather like today? Let's ask the children and see what they say!

Mittens: *(hold up mittens)* We'll wear our mittens out today! It looks like it's going to be a snowy day!

All: What's the weather like today? Let's ask the children and see what they say!

Umbrellas: *(hold up umbrellas)* We'll bring our umbrellas out today! It looks like it's going to be a rainy day!

All: What's the weather like today? Let's ask the children and see what they say!

All: We don't care! Let's go out and play! No matter the weather we'll have fun today!

Weather Items

Shared Writing

Umbrella Word Wall

MATERIALS

- ✓ red butcher paper
- ✓ construction paper
- ✓ silver glitter
- ✓ chart paper
- ✓ fishing line
- ✓ Fun Foam
- ✓ wooden dowel
- ✓ curling ribbon

Cut out a large umbrella from red butcher paper. (You may want to double the butcher paper for extra durability.) Cut out large raindrops from light blue construction paper, and outline them with silver glitter. Type the title *Our Raindrop Word Wall*, cut it out, and glue it on the umbrella. Invite the class to brainstorm weather words. Write each weather word on a raindrop, and use fishing line to attach the raindrops to the umbrella. Reread the word bank words with the class for a shared reading experience. Make a rainbow out of Fun Foam, glue it to a dowel, and attach curling ribbon in a rainbow of colors to make a fun reading stick. Invite children to use the rainbow reading stick to point to the words as they read them. Encourage children to refer to the word wall during independent writing.

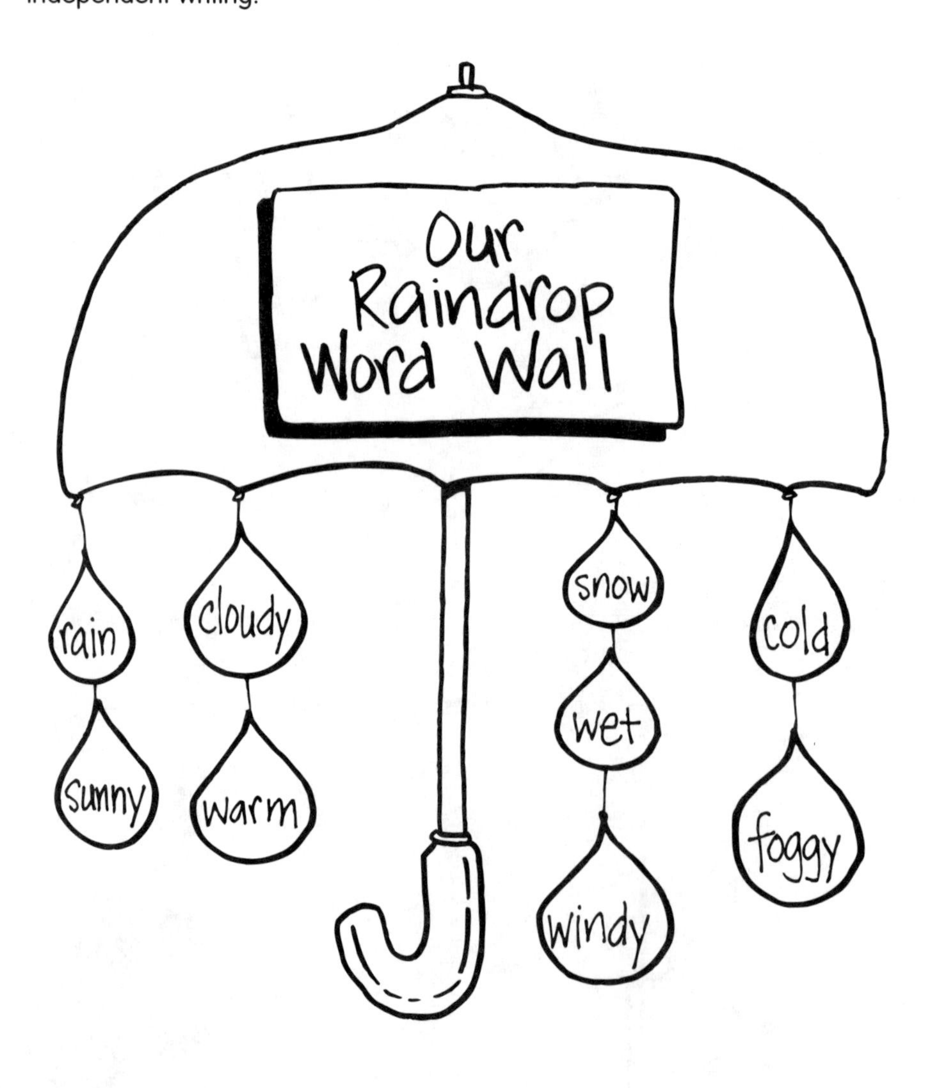

Weather Wizard of the Day

MATERIALS

- ✓ Weather Wizard reproducible (page 32)
- ✓ hat and cape
- ✓ photo of each child wearing the wizard hat
- ✓ butcher paper

In advance, take a picture of each child wearing a wizard hat. Cut butcher paper into the shape of a wizard hat to create pages for a class book. Make the pages large, and include one for each child. Invite each child to be the Weather Wizard of the Day. Have the Weather Wizard wear the wizard hat and cape and sit in a special chair. Invite the class to sing the following verse to the tune of "London Bridge":

What's the weather like today, like today, like today?
What's the weather like today? Let's ask the Weather Wizard!

Have the Weather Wizard go outside to check the weather. Have him or her give the weather report by singing the following verse to the tune of "London Bridge":

*It is a **rainy** day! A **rainy** day! A **rainy** day!*
*It is a **rainy** day! I am the Weather Wizard!*

Write the Weather Wizard's response on a Weather Wizard reproducible. Glue the sentence frame and that child's picture to a page in the class book. Each day, begin the activity by rereading the previous days' pages for shared reading. To extend the activity, make several copies of the Weather Graph (page 33), hole-punch each page, and insert the pages in a three-ring binder labeled *What Is the Weather Today?* Make sure there is one page for each month of the school year. Have the Weather Wizard record the weather for the day by coloring a square on the graph for that type of weather.

Weather Wizard

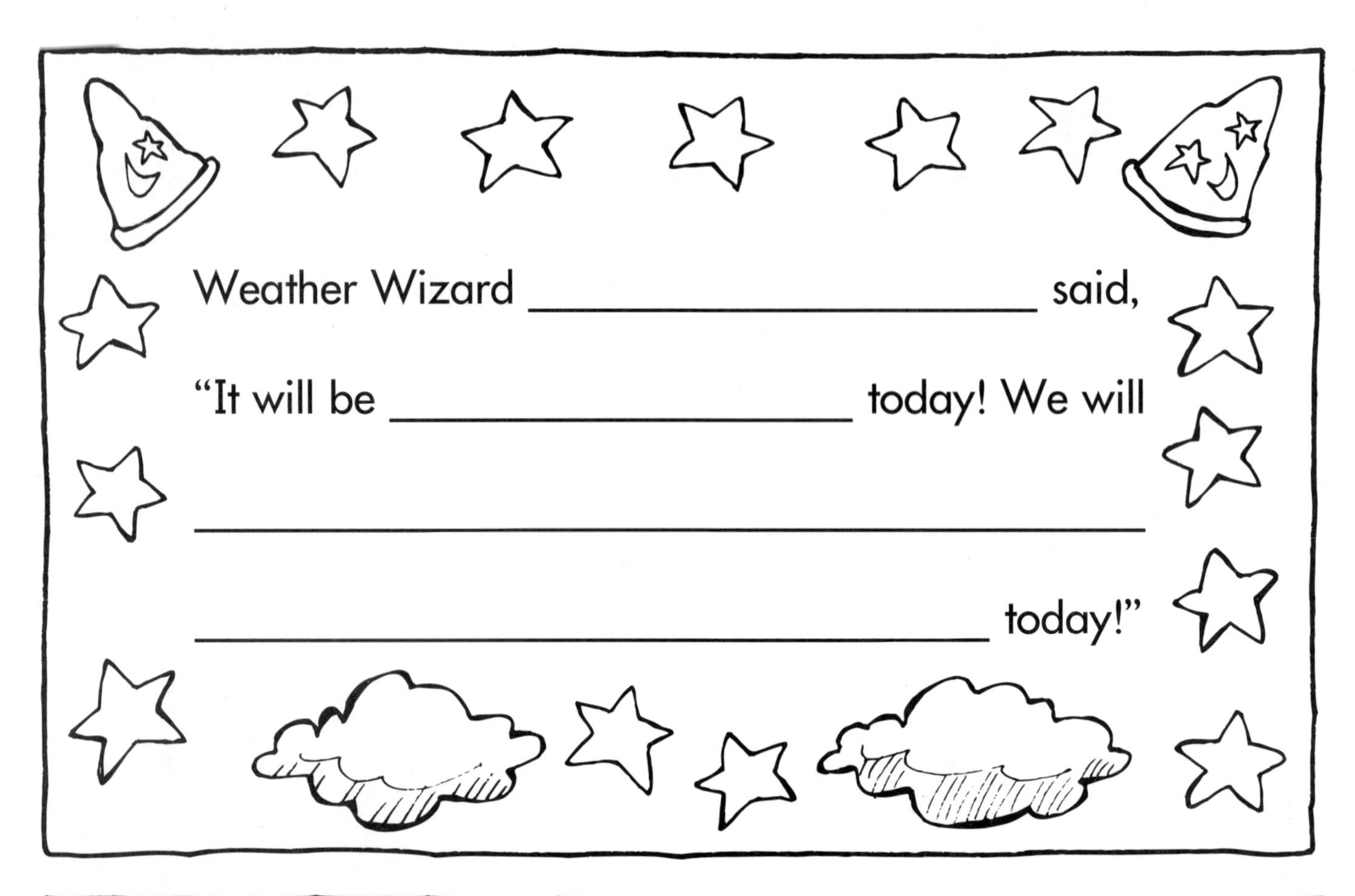

Weather Graph

for the month of ______________________________

sunny										
windy										
rainy										
overcast										
foggy										
snowy										

Interactive Writing

Weather Facts

MATERIALS

- ✓ register tape or construction paper strips
- ✓ butcher paper
- ✓ construction paper
- ✓ art supplies
- ✓ glitter star
- ✓ dowel

Have children discuss weather-related facts such as what a meteorologist does, types of clouds, and information about wind, rain, or other types of weather. Use interactive writing to record the information on register tape or construction paper strips. Help the class create a life-size Weather Wizard (a wizard who looks like Merlin with a gray beard, tall hat, and robe with stars). Invite children to paint the wizard on butcher paper or make him out of construction paper shapes. Glue children's sentences onto the wizard, and hang it in the classroom. Attach a glitter star to a dowel to create a reading wand. Invite children to use the wand to point to the words as they read the sentences.

Where Is Little Mouse?

MATERIALS

- ✓ *Mousekin's Golden House* by Edna Miller (Simon & Schuster)
- ✓ register tape, sentence strips, or construction paper strips
- ✓ black butcher paper
- ✓ construction paper

Read aloud *Mousekin's Golden House*, and discuss with the class why Mousekin seeks shelter in the pumpkin. Use interactive writing to write the story shown below. Print the story on register tape, sentence strips, or construction paper strips. Glue the sentences to five separate pieces of 18" x 24" (46 cm x 61 cm) black butcher paper. Divide the class into five small groups. Give each group one paper. Invite groups to tear construction paper into small pieces and glue them to their butcher paper to create a "tear art" picture to go with the sentence.

Where Is Little Mouse?

Page 1—Where is Little Mouse? He is in the leaves. It is fall!
Page 2—Where is Little Mouse? He is in the snow. It is winter!
Page 3—Where is Little Mouse? He is in the flowers. It is spring!
Page 4—Where is Little Mouse? He is in the pond. It is summer!
Page 5—Where is Little Mouse? (On the back of the page) *Here I am!*

Guided Writing

Weather Warm-Up

MATERIALS

- ✓ Mug reproducible (page 36)
- ✓ hot chocolate mix
- ✓ mugs
- ✓ marshmallows
- ✓ construction paper
- ✓ cotton balls
- ✓ close-up photo of each child's face
- ✓ art supplies

Discuss with the class how they can warm up when it is cold outside. Make hot chocolate, and invite children to have a cup. Ask children to decide if they like marshmallows and add them to their drink if they do. After children finish drinking their hot cocoa, write on the board *I ____________ marshmallows in my cocoa!* Model how to complete the frame by writing *I do not like marshmallows in my cocoa!* and *I like marshmallows in my cocoa!* Give each child a Mug reproducible. Ask children to complete the sentence frame by writing *like* or *do not like* on the blank line. Have children color the mug and cut it out. Ask them to also cut out their sentence frame. Have children glue their mug and sentence frame to a piece of construction paper. Invite children to glue cotton balls above their mug to represent marshmallows if they liked them. Ask children to glue their photo next to their mug and add a construction paper winter cap to their head. Invite children to trace their hands on tan or brown construction paper, cut out their tracings, and glue them on the mug so it looks like they are holding it. Display children's pictures and writing in the classroom, or bind them together into a class book. To extend the activity, make a large butcher paper mug, and glue three mugs at the bottom: one with hot cocoa in it, one with marshmallows in it, and one that is empty. Invite children to glue their photo above a cup to show whether they liked plain hot cocoa, liked hot cocoa with marshmallows, or did not like hot cocoa. Have the class analyze the data in the bar graph, and use interactive writing to tell about it (e.g., 4 like hot cocoa, 10 like marshmallows, 3 did not like hot cocoa).

Mug

I ______________________________ marshmallows

in my cocoa!

Independent Writing

The Snowy Day

MATERIALS

- ✓ *The Snowy Day* by Ezra Jack Keats
- ✓ chart paper
- ✓ construction paper
- ✓ thin black markers

Read aloud *The Snowy Day*, and discuss with the class the different activities children can do in the snow. Record children's ideas on chart paper (e.g., Chelsea can make snow angels, Bailey can sled in the snow). Have children independently write a sentence about an activity they can do in the snow. Encourage children to refer to word banks and use "temporary spelling" when they write. Give each child a 9" (23 cm) square of turquoise construction paper. Invite children to tear white construction paper and glue their pieces to the turquoise paper to create a snowy ground background. Then, invite them to tear a body, hat, and arms from red construction paper. Ask children to tear a face from tan or brown construction paper, mittens from yellow construction paper, and boots from black construction paper. Have children glue these pieces on their turquoise paper. Ask children to use a thin black marker to draw their face. Glue children's writing to the bottom of their picture. Display children's pictures and writing on a winter bulletin board surrounded by snowflakes. Invite children to share what they like to do in the snow.

Clouds

MATERIALS

- ✓ *It Looked Like Spilt Milk* by Charles G. Shaw or *Little Cloud* by Eric Carle
- ✓ white butcher paper
- ✓ blue construction paper
- ✓ white paint
- ✓ fishing line

Read aloud *It Looked Like Spilt Milk* or *Little Cloud.* Discuss with the class the various types of clouds in the sky and how clouds can give a clue as to what the weather may be that day. Tell children that clouds form when tiny drops of water vapor rise into the air; we cannot see these little droplets of water, but when they start collecting together, we see them as clouds. Explain that there are three basic types of clouds (cirrus, stratus, and cumulus). Tell children that cirrus clouds are white wispy clouds high up in the sky, stratus clouds are low clouds that sometimes give us rain, and cumulus clouds are puffy clouds that change form. (Cumulus clouds are the types of clouds talked about in the read-aloud books. They quickly change shape and often look like animals and faces. These clouds are visible on sunny days.) Give each child a piece of blue construction paper, and ask children to fold their paper in half and then reopen it. Squirt a blob of white paint on each child's paper. (Squirt it in a different spot on each child's paper.) Tell children to refold their paper and rub it to spread the paint around to make a "cloud." Have children open their paper and set it aside to dry. Invite children to look at their cloud the next day to see what it looks like to them. Have children write about what their cloud looks like on a separate piece of paper. Have them use the following frame as a guide for their writing: *"My cloud looks like a ________!" said* ________. Cut each child's blue paper into a cloud shape. Cut out larger clouds from white butcher paper, and attach each blue cloud shape to a separate white cloud. Then, attach each child's writing to his or her cloud. Display the completed clouds on fishing line, or combine them all to create a class book titled *Little Cloud by the Cumulus Clouds in Room ____!*

Have You Ever Built a Snowman?

MATERIALS

- ✓ *The Biggest, Best Snowman* by Margery Cuyler
- ✓ chart paper
- ✓ markers

Read aloud *The Biggest, Best Snowman.* Write at the top of a piece of chart paper *Have you ever built a snowman?* Read aloud the question, and model how to write the answer in a complete sentence. For example, write *"Yes, I have," said Mrs. Corcoran* or *"No, I have not," said Mrs. Corcoran.* During circle time, center time, or writing time, invite children to use your answer as a frame to write their own sentence on the chart paper. (It may take several days for all the children to have an opportunity to write on the group chart.) Remind children to refer to word banks and environmental print and use "temporary spelling" when they write their answer. After all of the children have written their answer, invite the class to read with you the question and each child's answer. To extend the activity, write a different question each week and have children write an answer. Hang each week's question and answers on a wall or chart rack, and encourage children to reread the pages for independent reading.

Have you ever built a snowman?

"Yes, I have," said Mrs. Corcoran.

"No, I have not," said Joey.

"Yes, I have," said Kim.

Math

Edible Rainbows

MATERIALS

- ✓ Rainbow Math reproducible (page 43)
- ✓ paper or plastic cups
- ✓ colorful treats (e.g., jelly beans, Skittles®, M&M's®, Froot Loops® cereal)

Give each child a cup of assorted colorful treats and a Rainbow Math reproducible. Invite children to sort their treats by color on the rainbow arches of the reproducible. Have children count how many of each color they have and write the number on the corresponding arch. To extend the activity, have children fill in the frames below the rainbow to create addition or subtraction equations. For example, if a child has 7 green treats, and 5 red treats, he or she could write $7 + 5 = 12$ or $12 - 5 = 7$.

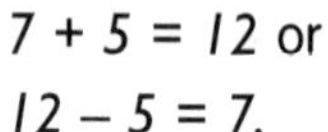

Rainbow Colors

MATERIALS

- ✓ Rainbow Data Sheet (page 44)
- ✓ butcher paper
- ✓ newspaper or paper towels
- ✓ 3" x 2½' (7.5 cm x 76 cm) strips of construction paper (1 strip of each color: red, orange, yellow, green, blue, purple)
- ✓ close-up photo of each child or 2" (5 cm) white paper squares

Cut out two large identical clouds from white butcher paper. Glue together the edges of the clouds, but leave a small opening to stuff the cloud with newspaper or paper towels. Then, glue the opening to "close" the cloud. Attach the colored strips of construction paper to the bottom of the cloud. Invite the class to make a graph of the children's favorite rainbow colors. Have children glue their photo to the strip of their favorite rainbow color, or have them draw a picture of their head on a white paper square and glue it to the strip. Then, invite the class to analyze the data on the graph by completing the Rainbow Data Sheet. Glue the data sheet on the back side of the cloud. Write *Rainbow colors—oh so fun! / Which color is your favorite one?* on the front side, and hang the cloud from the ceiling.

Seasons Glyph Quilt

MATERIALS

- ✓ books about seasons
- ✓ Seasons Glyph Key reproducible (page 45)
- ✓ Clothing reproducible (page 46)
- ✓ construction paper
- ✓ ribbon (optional)
- ✓ art supplies

Discuss with the class the four seasons and how the weather typically changes during each season. Cut out 9" (23 cm) construction paper squares (two for each child). Cut light blue squares for boys, violet for girls, and dark blue for the season pictures. Give each child a dark blue square. Have children tear brown construction paper and glue their torn pieces to their square to create a "tear art" tree trunk and branches. Invite children to sponge-paint their tree to match their favorite season. For example, green leaves with flowers for spring, snow-topped branches for winter, fall-colored leaves for fall, and bright green leaves for summer. Give each girl a violet square and each boy a light blue square. Have children choose the item from the Clothing reproducible that represents their favorite season. Ask them to trace the clothing item on their favorite color of construction paper, cut out the item, and glue it on their light blue or violet square. Next, have children add stripes to their clothing to represent how old they are (e.g., 6 stripes if they are 6 years old). Ask children to write their name on both squares. Arrange the completed squares on a bulletin board. Alternate each tree and clothing square. Tie ribbons in between the squares to make the display look like a quilt. Or, hole-punch children's squares and tie them together with ribbon bows to make a "quilt." Display the quilt, and have children use the Seasons Glyph Key to analyze the data on each square. For example, a child might say *Chelsey is a girl because she has a violet square. She likes summer because her tree has green leaves and she has a bathing suit on her square. Her bathing suit is pink because pink is her favorite color. She is 6 years old because there are 6 stripes on her bathing suit.*

Winter Ice Cube Counting

MATERIALS

- ✓ Number Grid (page 47)
- ✓ black permanent marker
- ✓ ice cube trays
- ✓ small snowmen or winter animal erasers

Use a black permanent marker to write the numbers 1–10 in random order in the bottom of ice cube trays. (Write one number in the bottom of each section.) Give each child a tray, a Number Grid, and an eraser. Invite children to toss the eraser into the tray and write the number in the matching column on the grid. For example, if the eraser lands on the number 6, the child writes *6* in the 6 column of the grid. Have children repeat this process ten times. Ask children which number came up most often. To extend the activity, use larger numbers or write addition or subtraction problems with sums that are ten or less in the bottom of the trays. Have children write the sum in the correct column of the grid.

Name ____________________

Rainbow Math

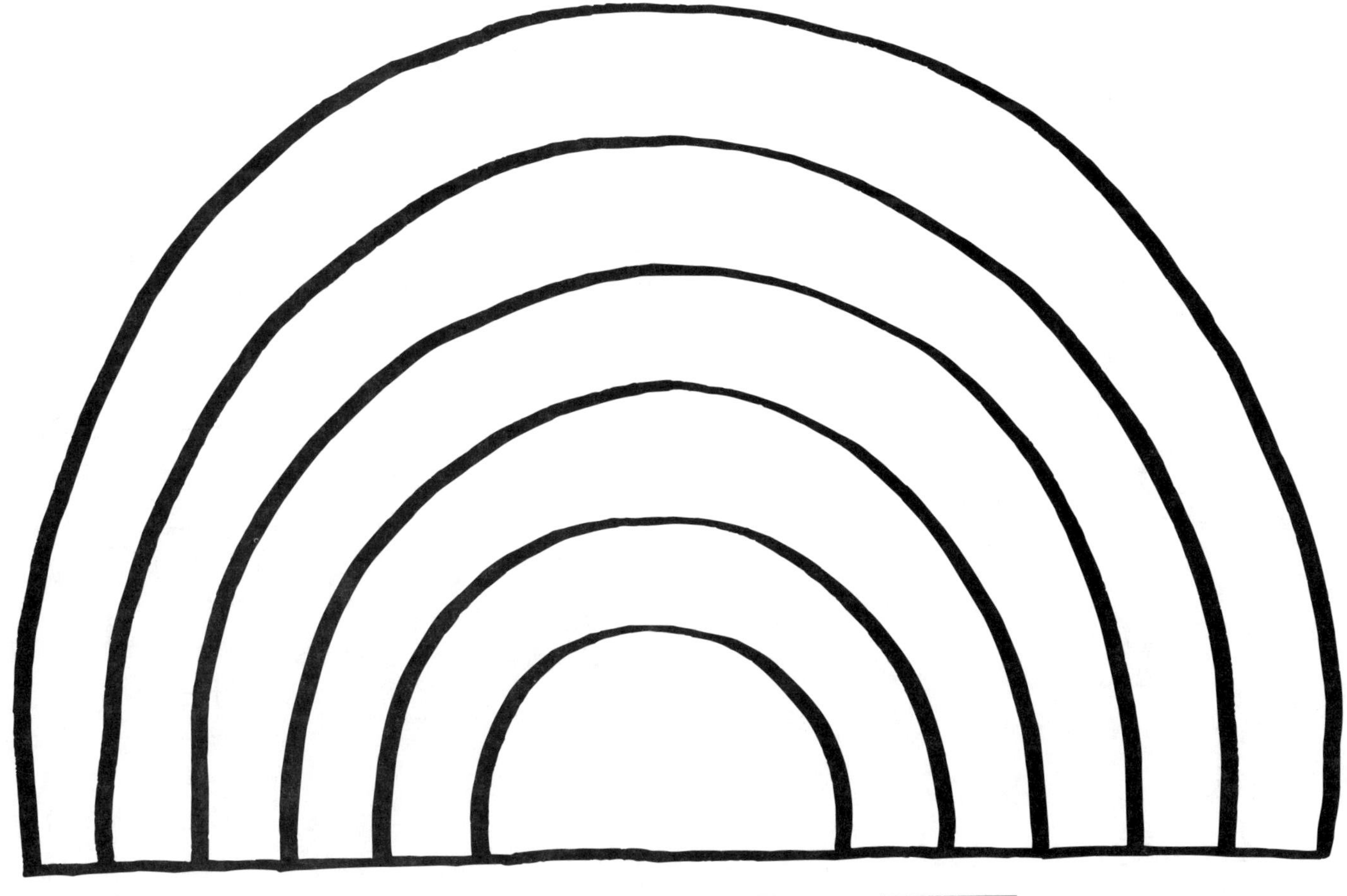

☐ ___ ☐ = ☐

☐ ___ ☐ = ☐

☐ ___ ☐ = ☐

☐ ___ ☐ = ☐

Rainbow Data Sheet

What colors were the favorites?

______________ and ______________

More children liked ______________

than ______________.

Less children liked ______________

than ______________!

___________ children participated

in this graph!

Seasons Glyph Key

1. What is your favorite season?
   Spring: Sponge-paint green leaves with flowers on your tree.

   Winter: Sponge-paint snow on the branches on your tree.

   Fall: Sponge-paint colored leaves on your tree.

   Summer: Sponge-paint bright green leaves on your tree.

2. Are you a boy or a girl?
   If you are a boy, use a light blue square.

   If you are a girl, use a violet square.

3. What is your favorite color?
   Trace the clothing on a piece of paper in your favorite color.

4. How old are you?
   Make the same number of stripes on the clothing as your age.

Clothing

Name ____________________

Number Grid

1	2	3	4	5	6	7	8	9	10

Science

Sunshine Solar Sandwiches

MATERIALS

- ✓ pie pans
- ✓ graham crackers
- ✓ mini-marshmallows
- ✓ mini-M&M's®
- ✓ aluminum foil
- ✓ permanent markers

Divide the class into pairs, and give each pair a pie pan. Have each child place a graham cracker in the pan and sprinkle mini-marshmallows and mini-M&M's on top of the graham cracker. Cover each pan with aluminum foil, and use a permanent marker to write both children's names on the foil. Place children's pans outside in the sun. Have children check their pan every 10 to 15 minutes to see what is happening. Once children notice that their ingredients have melted, invite them to add another graham cracker on top and eat their "solar sandwich." Discuss with the class why the chocolate and marshmallows melted.

Rainy Day in a Bag

MATERIALS

- ✓ several books about rain and the water cycle
- ✓ Rain in a Bag Frame (page 50)
- ✓ sponges
- ✓ resealable plastic bags

Copy a class set of the Rain in a Bag Frame, and cut out the center of each one. Read aloud several books about rain and the water cycle such as *Thunder Cake* by Patricia Polacco. Give each child one fourth of a sponge, a plastic bag, and a Rain in a Bag Frame. Have children wet their sponge, place it in the plastic bag, and seal the bag. Invite children to decorate their frame and tape it to the front of their bag. Tape the bags to a sunny window. Ask children to predict what will happen. After a few hours, have children observe what happened (the bag will have little drops of condensation). Tell children that clouds are made of little drops of water like those in the bags and that cool air in the sky makes rain fall out of real clouds. Discuss how children similarly made "rain" fall from their "clouds." Explain that the moist air near the earth's surface (the sponge) rises and hits the cooler air in the clouds, condenses, and becomes rain!

The Wind Is Blowing

MATERIALS

- ✓ *Chicka Chicka Boom Boom* by Bill Martin Jr. and John Archambault (Simon & Schuster)
- ✓ *The Tiny Seed* by Eric Carle
- ✓ Pinwheel reproducible (page 51)
- ✓ construction paper
- ✓ orange straws
- ✓ brass fasteners
- ✓ alphabet stickers

Copy a class set of the Pinwheel reproducible on green construction paper. Read aloud *Chicka Chicka Boom Boom* (alphabet book) and *The Tiny Seed* (book about the wind). Discuss how wind makes things move (e.g., sails on a boat, kites fly, whistles blow, wind helps distribute seeds). Give each child a Pinwheel reproducible and an orange straw, and invite children to make a pinwheel. Have them cut out the pinwheel and cut on the dotted lines. Use a hole punch to punch a hole in each child's straw and the circles on the pinwheel. Have children fold each corner of the pinwheel onto the center hole and use a brass fastener to hold the corners in place. Invite children to add alphabet stickers to the green part of the pinwheel to make an "ABC coconut tree" (like in the book *Chicka Chicka Boom Boom*). Have children cut out little brown circles for "coconuts" and glue them to the pinwheel to add a fun look. Have children test out their pinwheel. Tell them to hold it still and see if it moves. Then, invite children to take their pinwheel outside to see if it will move. Ask children *Is there a breeze? Does your pinwheel spin?* Ask children to blow on their pinwheel to see if it will spin. Ask children what scientific information they have learned, and chart their responses.

Rain in a Bag Frame

Pinwheel

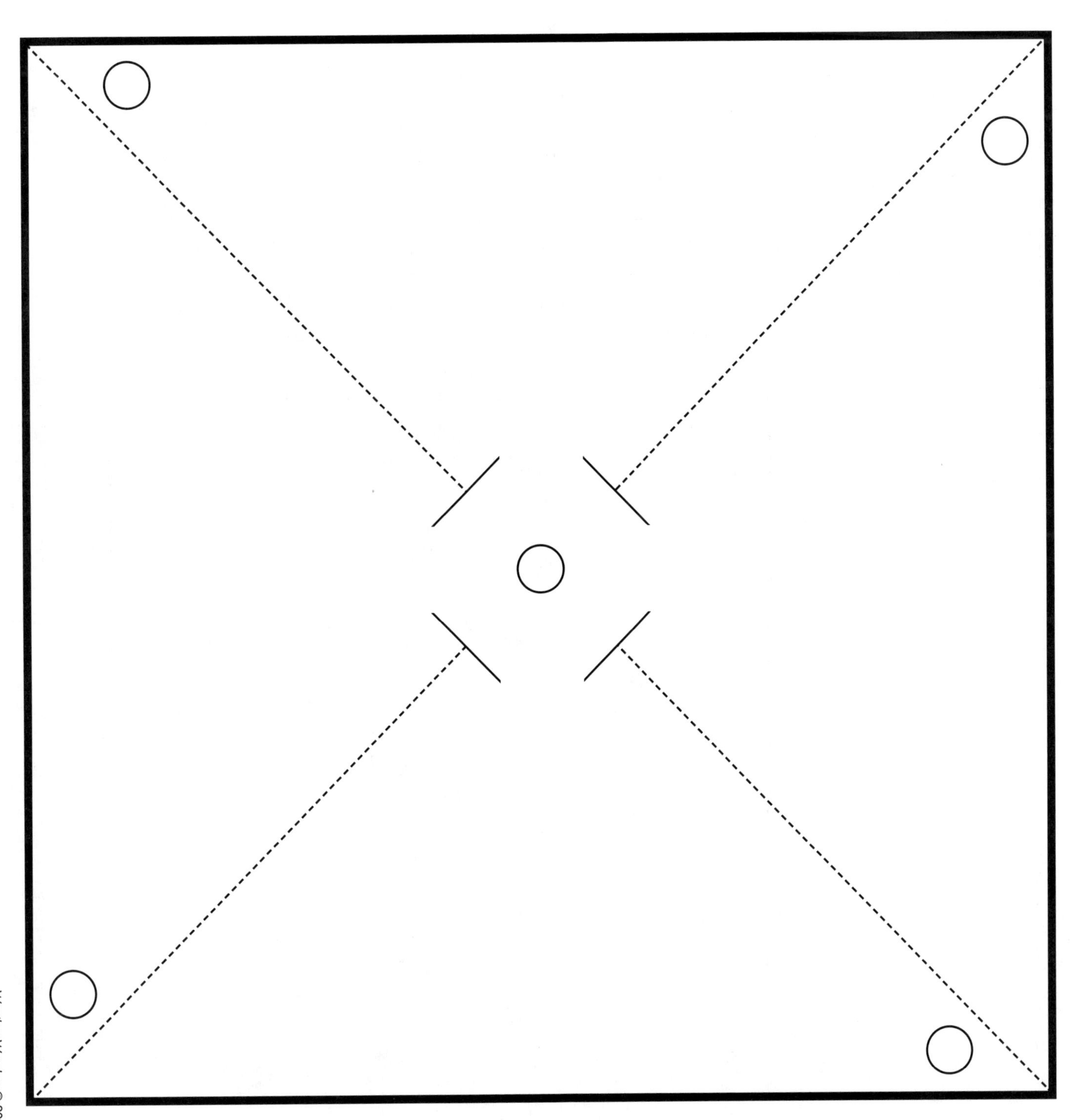

Social Studies

My Big Yellow Boots

MATERIALS

- ✓ several books about manners
- ✓ Big Yellow Boots reproducible (page 54)
- ✓ construction paper
- ✓ art supplies

Copy a class set of the Big Yellow Boots reproducible on yellow construction paper. Read aloud books about manners such as *Big Black Bear* by Wong Herbert Yee (Houghton Mifflin). Discuss manners and what children should do around puddles on rainy days. Ask children why someone might be upset if he or she were splashed. Invite children to paint a self-portrait of themselves on a rainy day. Give each child a reproducible, and have children cut out the boots. Have them complete the sentence frame on their reproducible by writing what their boots can do (e.g., splash, jump) in the first blank and a person's name (e.g., Mom, Dad) in the second blank. Have them glue the boots to the feet of their self-portrait. Combine children's completed pages to create a class book. Encourage children to read it for independent reading.

Feelings for Weather

MATERIALS

- ✓ several books about weather
- ✓ Paper Doll reproducible (page 55)
- ✓ Paper Doll Clothes reproducible (page 56)
- ✓ construction paper
- ✓ white butcher paper
- ✓ art supplies

Copy a class set of the Paper Doll and Paper Doll Clothes reproducibles on construction paper. Read aloud several books about weather such as *Weather Words and What They Mean* by Gail Gibbons, and discuss with the class the different types of weather. Have children pick their favorite kind of weather and say how it makes them feel. Give each child a copy of the two reproducibles. Invite children to cut out their paper doll and dress it up to reflect their weather choice. Have each child paint a background scene that shows his or her weather choice on an 18" x 24" (46 cm x 61 cm) piece of white butcher paper. When the paint is dry, attach children's completed paper doll to their picture. Have children complete the sentence frame _________ *feels* ______. Label each picture with the weather word. Display the pictures in your classroom, or combine children's finished pages to create a class Big Book titled *Feelings for Weather.*

Big Yellow Boots

My big yellow boots can

______________________ in puddles but not on

__________________________ or I'll be in trouble!

(name)

Paper Doll

Paper Doll Clothes

Culminating Event and Extra Fun

At the end of your unit, host a Weather Wonderland celebration. Invite children and their families to visit the classroom so children can act as a guide to show off all the projects they completed during the unit and share the information they learned about weather. Arrange your classroom so all the projects children completed are displayed. Prior to the event, have children practice leading a partner or small group around the classroom and explaining each project. This will help prepare children and make them feel confident when their family visits the classroom. Invite children to complete the following fun activities to provide decorations and props for the "big event."

Weather Wonderland Invitation

MATERIALS

- ✓ Weather Wonderland Invitation (page 61)
- ✓ construction paper
- ✓ glitter

Ask children to make a special invitation to invite family members to the Weather Wonderland celebration to see what their little Weather Wizards have been learning. Have children make a wizard hat by folding purple construction paper to make a triangle. Invite them to add yellow stars and glitter to decorate their hat. Give each child a Weather Wonderland Invitation to complete. Ask children to glue their completed invitation to the front of their wizard hat.

Snow Globes

MATERIALS

- ✓ baby food jars
- ✓ glue gun
- ✓ small plastic animals
- ✓ glitter or fake snow
- ✓ ribbon, holiday greenery, or food coloring

Give each child a clean baby food jar. Have children choose a small plastic animal. Hot-glue children's animals on the underside of their jar lid. When the glue is dry, add cold water to the jar (fill to within a ½" or 13 mm of the top). Add 1 tablespoon (15 mL) of glitter, fake snow, or both to the jar. Use hot glue to seal the jar. Have children decorate their jar with ribbon or holiday greenery, or add food coloring to the water. Use your imagination, and have fun!

Create a Weather Wonderland

MATERIALS

- ✓ playhouse
- ✓ teacher-made tree
- ✓ construction paper
- ✓ white batting
- ✓ tissue paper
- ✓ fishing line
- ✓ beach towels, buckets, sunglasses, beach balls
- ✓ art supplies

Decorate a classroom playhouse and tree according to the weather of each season. In **fall,** invite children to cut out leaves and sponge-paint them with fall colors to display on the playhouse and in the tree. In **winter,** invite children to make snowflakes out of white construction paper and hang them from the tree. Place white batting on the top of the playhouse for a snowy effect. In **spring,** invite children to make tissue paper butterflies and flowers to decorate your playhouse. Use fishing line to hang the butterflies from the tree. In **summer,** add beach towels, sand buckets, and sunglasses to your playhouse. Hang beach balls from your tree for a fun, sunny summer look. Ask parents to make a tablecloth out of material for your playhouse table to match the weather of each season (e.g., leaves and pumpkins for fall, umbrellas and raindrops for spring). You can buy inexpensive paper plates and napkins for children to use in the playhouse to correspond with the season.

Cloud Viewers

MATERIALS

- ✓ blue file folders cut in half or blue tagboard (8½" x 11" or 21.5 cm x 28 cm)
- ✓ cotton

Discuss with the class various types of clouds and how they take on many forms (e.g., faces, plants, animals). Give each child a blue file folder or tagboard, and have children cut a cloud-shaped hole in the center of it. Tell them to glue cotton all around the edges to make it look like clouds. (Tell children to stretch the cotton so they do not need to use as much.) When the glue is dry, invite children to go outside, lay down on the grass, and look for clouds in their "cloud viewer." Encourage them to use their imagination to see what shapes they can find.

Windy Day

M A T E R I A L S

- ✓ *The Wind Blew* by Pat Hutchins
- ✓ Cloud reproducible (page 62)
- ✓ X-ACTO® knife
- ✓ sturdy large round dinner paper plates and soup bowls
- ✓ white butcher paper
- ✓ newspaper
- ✓ art supplies

Read aloud *The Wind Blew.* Discuss with the class how the wind can be very strong at times. Use an X-ACTO knife to cut a hole in the center of a large paper plate for each child. Glue a paper bowl to each plate to create "hats." Give a hat to each child, and have children paint their hat in bright colors. Invite them to decorate their hat with glitter, feathers, sequins, and/or macaroni. Create a fun bulletin board display. Make a large stuffed butcher paper cloud. Add a face blowing like the wind. Attach the cloud to one corner of a bulletin board. Add the following text to the display: *Hold on to your hat, / Wherever you are, / Or the wild wind will blow it far!* Attach all of the hats to the bulletin board in a haphazard way so they look like the wind is blowing them. Give each child a Cloud reproducible, and have children write where their hat is being blown to (e.g., the zoo, the moon). Display children's writing on the bulletin board.

Springtime Rainbow Bubbles

MATERIALS

- ✓ water table or buckets
- ✓ Dawn® dishwashing liquid
- ✓ food coloring
- ✓ tools (e.g., flyswatters, straws, eggbeaters, whisks, cookie cutters)
- ✓ waterproof smocks (optional)

Fill your water table (outside) or buckets with two parts water to one part clear Dawn dishwashing liquid. Add a few drops of food coloring. Have children explore making bubbles with a variety of tools. Have children wear a waterproof smock, or do this activity on a warm spring day. As children make bubbles, ask them if they can find a rainbow in their bubbles. Ask them what colors they see. Try this activity on a windy day to see how far their bubbles will travel.

Bubble Paint

MATERIALS

- ✓ bowls
- ✓ 1/2 cup (125 mL) water
- ✓ 1/4 cup (50 mL) Dawn® dishwashing liquid
- ✓ 2 teaspoons (10 mL) tempera paint
- ✓ straws
- ✓ white construction paper

Make three to four bowls of colored "bubble water" by mixing together water, dishwashing soap, and tempera paint. Invite children use a straw to blow in the bubble water. When there are bubbles floating at the top of the bowl, have children gently pat the bubbles on a piece of white construction paper. The bubbles will pop and make fun bubble art! These designs will make fun book covers, journal covers, or picture frames.

Cooking

MATERIALS

- ✓ Sunshine Rays Recipe (page 63)
- ✓ Edible Snowflake Recipe (page 64)

Gather for each child the ingredients and materials listed on a recipe. Give each child a copy of the recipe. Have children follow the recipe and color each illustration. For the Sunshine Rays Recipe, mix the cream cheese with yellow food coloring to make yellow cream cheese. For the Edible Snowflake Recipe, warm the tortillas prior to the activity.

Weather Wonderland Invitation

Dear Family,

We have been very busy learning about different types of weather! Please join us for our Weather Wonderland celebration!

Where: ______________________

When: ______________________

Time: ______________________

I have so much to show you! Please dress according to the weather!

Love,
Weather Wizard

____

Cloud

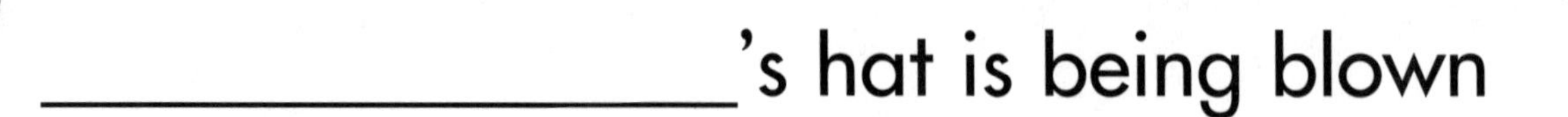

________________'s hat is being blown

all the way to________________________!

________________'s hat is being blown

all the way to________________________!

Sunshine Rays Recipe

by Chef ______________________________

<u>Ingredients</u>: rice cake, cream cheese, yellow food coloring, pretzel sticks, orange or yellow sprinkles

<u>Materials</u>: plastic knife, napkin

1. Start with a round rice cake.

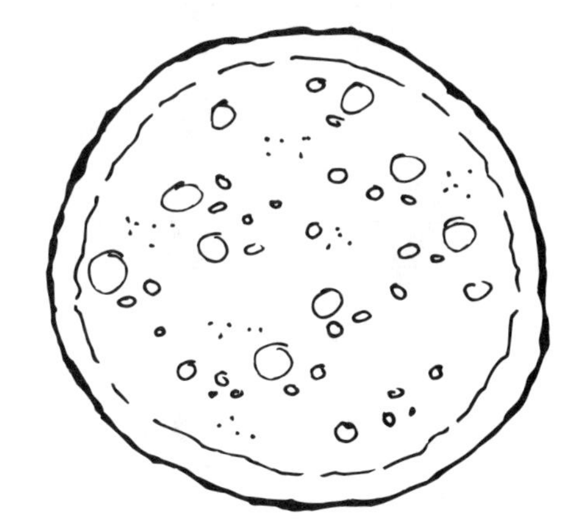

2. Spread yellow cream cheese on the rice cake.

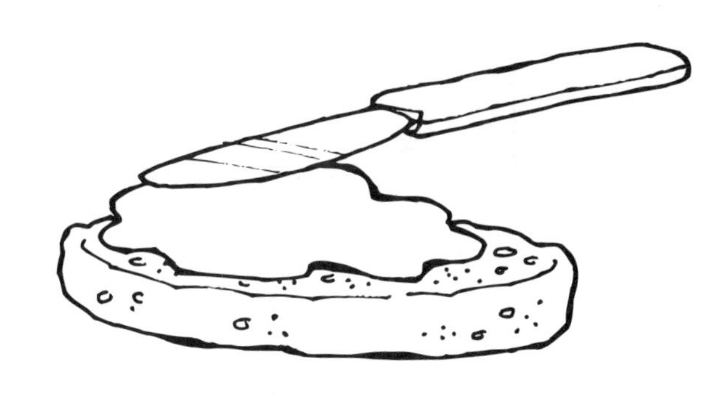

3. Add 8 pretzel sticks to make the sun's rays.

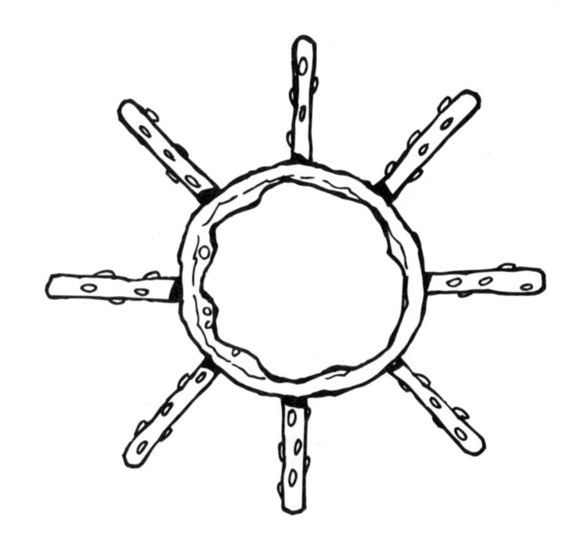

4. Shake colorful sprinkles on your sun. Enjoy your taste of sunshine!

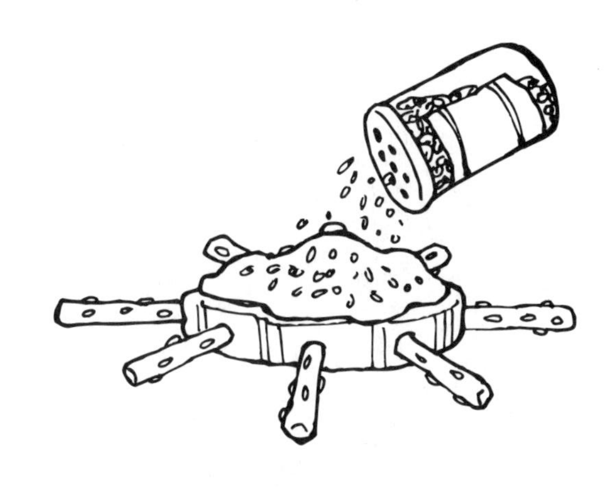

Edible Snowflake Recipe

by Chef ______________________

<u>Ingredients</u>: flour tortilla, melted butter, sugar

<u>Materials</u>: napkin

1. Fold a warm tortilla in half and then in half again.

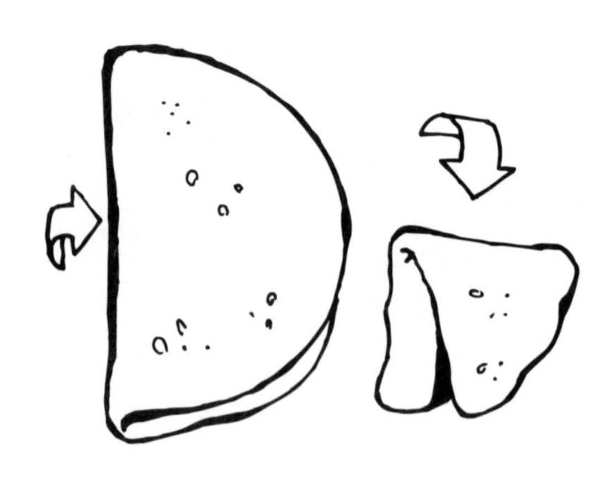

2. Take tiny bites along the edges of your tortilla.

3. Open up the tortilla and you will have a snowflake!

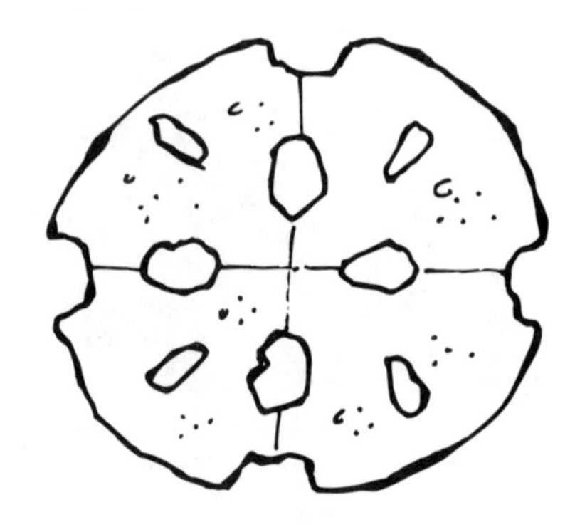

4. Add melted butter and sprinkle on sugar. Eat it up. Yum!